MEDIA GUIDE

Essentials of Gene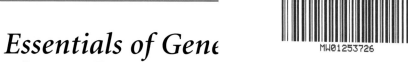
SECOND EDITION, WITH VIRTUAL TOOLBOX

Darrell D. Ebbing
Steven D. Gammon
Ronald O. Ragsdale

Virtual Toolbox
Flexible, integrated learning tools

This handy guide provides information about the media resources available with the Second Edition. These tools can help you succeed in your course.

Contents

SMARTHINKING™ live, online tutoring	4
Student Website	6
Eduspace® (powered by Blackboard™) (Please note, you can only access Eduspace® if your instructor has chosen to use it.)	8
Periodic Table of the Elements	10
Table of Atomic Masses	11
Print Supplements	12
System Requirements	13

Copyright © 2006 by Houghton Mifflin Company. All rights reserved.

No part of this work may be reproduced or transmitted in any form or by any means, electronic or mechanical, including photocopying and recording, or by any information storage or retrieval system without the prior written permission of Houghton Mifflin Company unless such copying is expressly permitted by federal copyright law. Address inquiries to College Permissions, Houghton Mifflin Company, 222 Berkeley Street, Boston, MA 02116-3764.

Printed in the U.S.A.

Media Guide for Students ISBN: 0-618-49247-X
Student Text ISBN: 0-618-49175-9
Student Text and Media Guide for Students ISBN: 0-618-57802-1

Visit: college.hmco.com/PIC/ebbingessentials2e   For Technical Support: (800) 732-3223 or support@hmco.com

Dear Student,

Welcome to your new adventure in chemistry! You will build on previous experiences with chemistry to develop a broader and deeper understanding of this marvelous and diverse field. Be aware, though, that the important skills you develop will include both the chemistry and the methods you develop to learn it. No one can *teach* you chemistry. You must *learn* the material on your own by using whatever resources are available, especially study aids and expert assistance.

Explore the Media Guide carefully to discover which of the many resources best suit your needs and method of study. The **textbook** is your first and most important resource and should always be your starting point to learn course content. The **Student Website** gives you the freedom to access additional study aids such as flashcards, a brief overview of each chapter, and a collection of molecular animations and lab demonstrations. It also offers a molecular library and ACE self-tests to help you prepare for quizzes and exams. In addition, you can use the Keywords list to test your understanding of each word or phrase or to clarify a new definition.

Through **SMARTHINKING**™, you can receive live, online tutoring from qualified instructors during peak study hours. In addition, you can submit questions and get feedback within 24 hours. **Eduspace**® provides another avenue to the resources described above, along with online homework assignments given by your instructor.

C. Weldon Mathews, Ohio State University

Visit: **college.hmco.com/PIC/ebbingessentials2e**   For Technical Support: **(800) 732-3223** or **support@hmco.com**

SMARTHINKING™ live, online tutoring

Why use SMARTHINKING?

- Access live help when you are unable to meet with your instructor.

- Get personalized tutoring with difficult concepts.

- Review previous problem-solving discussions to help prepare for exams.

This live, online service provides personalized, text-specific tutoring when you need it most.

With SMARTHINKING™ you can:

- connect immediately to live help during typical study hours: Sunday through Thursday from 2 P.M. to 5 P.M. and 9 P.M. through 1 A.M. EST.

- submit a question to get a response from a qualified e-structor within 24 hours.

- use the whiteboard with full scientific notation and graphics.

- pre-schedule time with an e-structor.

- view past online sessions, questions, or essays in an archive on your personal academic homepage.

- view your tutoring schedule.

E-structors help you with the process of problem-solving rather than supply answers.

▼ **Whiteboard**

▼ **SMARTHINKING™ live, online tutoring homepage**

Visit: www.smarthinking.com   For Technical Support: (888) 430-7429, ext. 1 or info@smarthinking.com

SMARTHINKING™ live, online tutoring

Logging on to SMARTHINKING live, online tutoring

1. Go to smarthinking.com/houghton.html.

2. Follow the instructions on-line to set up your student account using the password located on the inside front cover of this guide.

For technical support for SMARTHINKING, call (888) 430-7429, extension 1 or email info@smarthinking.com.

If your instructor asked you to register for *Eduspace,* you can also access *SMARTHINKING* directly from within *Eduspace* once you have logged in.

Visit: www.smarthinking.com   For Technical Support: (888) 430-7429, ext. 1 or info@smarthinking.com

Student Website *Essentials of General Chemistry,* 2/e

Why use the Student Website?

- Get help understanding core concepts at any time
- Visualize molecular-level interactions.
- Practice problem solving with ACE practice tests to prepare for quizzes and exams.

Organized by chapter, the *Student Website* supports the goals of the Second Edition with the following resources:

Visualization, Practice, and Tutorial Tools:

- Understanding Concepts and Visualizations tutorials that use images, animations, and videos to help reinforce core chemical concepts
- Quiz questions that test your knowledge of the Visualizations, animations, and videos
- Flashcards of key terms and concepts
- Houghton Mifflin's ACE practice tests
- Molecule library

Additional Resources and Study Aids:

- An interactive periodic table
- Glossary from the text
- Link to SMARTHINKING online tutoring
- Information on careers in chemistry

▼ Interactive flashcards

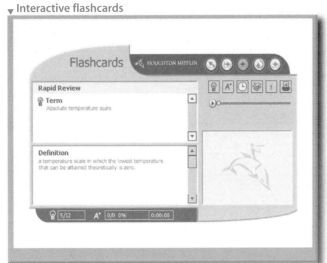

▼ ACE self-quizzes

Visit: **college.hmco.com/PIC/ebbingessentials2e**   For Technical Support: **(800) 732-3223** or **support@hmco.com**

Student Website *Essentials of General Chemistry, 2/e*

Logging on to the Student Website

1. **Go to college.hmco.com/ PIC/ebbingessentials2e.**
2. **Select Students.**
3. **Enter the user name and password located on the inside front cover of this guide.**

If your instructor asked you to register for *Eduspace*, you can also access the *Student Website* directly from within *Eduspace* once you have logged in.

Visit: college.hmco.com/PIC/ebbingessentials2e   For Technical Support: (800) 732-3223 or support@hmco.com

Eduspace® (powered by Blackboard™)

Why use Eduspace?

- Complete assignments specifically chosen by your instructor
- Practice problem solving with online homework to improve your scores on quizzes and exams
- Access live, online tutoring through SMARTHINKING anytime you need extra help
- Visualize chemical concepts at the molecular level
- Extend your understanding of chapter concepts

Online Homework and Gradebook

Accessible anytime you need it, *Eduspace* includes an automatically graded online homework system with three types of problems: practice exercises based on in-text examples with rejoinders, algorithmic end-of-chapter questions, and test bank questions. Using *Eduspace*, you can practice problem solving and check your progress in the gradebook at any time.

Online Homework

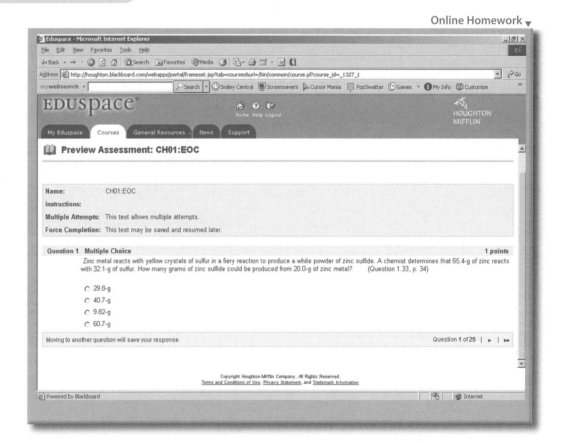

Visit: www.eduspace.com   For Technical Support: (866) 817-3139

Eduspace® (powered by Blackboard™)

Logging on to **Eduspace** (powered by Blackboard)

1. **Go to www.eduspace.com.**
2. **Register for *Eduspace*.**

If your instructor is using the *Eduspace* program, refer to your *Eduspace Getting Started Guide for Students* for information and access.

Visit: www.eduspace.com   For Technical Support: (866) 817-3139

Periodic Table of the Elements

For your reference

Group	1 (1A)	2 (2A)	3	4	5	6	7	8	9	10	11	12	13 (3A)	14 (4A)	15 (5A)	16 (6A)	17 (7A)	18 (8A)
1	1 H 1.008																	2 He 4.003
2	3 Li 6.941	4 Be 9.012											5 B 10.81	6 C 12.01	7 N 14.01	8 O 16.00	9 F 19.00	10 Ne 20.18
3	11 Na 22.99	12 Mg 24.31											13 Al 26.98	14 Si 28.09	15 P 30.97	16 S 32.07	17 Cl 35.45	18 Ar 39.95
4	19 K 39.10	20 Ca 40.08	21 Sc 44.96	22 Ti 47.88	23 V 50.94	24 Cr 52.00	25 Mn 54.94	26 Fe 55.85	27 Co 58.93	28 Ni 58.69	29 Cu 63.55	30 Zn 65.38	31 Ga 69.72	32 Ge 72.59	33 As 74.92	34 Se 78.96	35 Br 79.90	36 Kr 83.80
5	37 Rb 85.47	38 Sr 87.62	39 Y 88.91	40 Zr 91.22	41 Nb 92.91	42 Mo 95.94	43 Tc (98)	44 Ru 101.1	45 Rh 102.9	46 Pd 106.4	47 Ag 107.9	48 Cd 112.4	49 In 114.8	50 Sn 118.7	51 Sb 121.8	52 Te 127.6	53 I 126.9	54 Xe 131.3
6	55 Cs 132.9	56 Ba 137.3	57 La* 138.9	72 Hf 178.5	73 Ta 180.9	74 W 183.9	75 Re 186.2	76 Os 190.2	77 Ir 192.2	78 Pt 195.1	79 Au 197.0	80 Hg 200.6	81 Tl 204.4	82 Pb 207.2	83 Bi 209.0	84 Po (209)	85 At (210)	86 Rn (222)
7	87 Fr (223)	88 Ra 226	89 Ac† (227)	104 Rf (261)	105 Db (262)	106 Sg (263)	107 Bh (264)	108 Hs (265)	109 Mt (268)	110 Ds (281)	111 Uuu	112 Uub		114 Uuq				

*Lanthanides:

58 Ce 140.1	59 Pr 140.9	60 Nd 144.2	61 Pm (145)	62 Sm 150.4	63 Eu 152.0	64 Gd 157.3	65 Tb 158.9	66 Dy 162.5	67 Ho 164.9	68 Er 167.3	69 Tm 168.9	70 Yb 173.0	71 Lu 175.0

†Actinides:

90 Th 232.0	91 Pa (231)	92 U 238.0	93 Np (237)	94 Pu (244)	95 Am (243)	96 Cm (247)	97 Bk (247)	98 Cf (251)	99 Es (252)	100 Fm (257)	101 Md (258)	102 No (259)	103 Lr (260)

Group numbers 1–18 represent the system recommended by the International Union of Pure and Applied Chemistry.

Visit: college.hmco.com/PIC/ebbingessentials2e   For Technical Support: (800) 732-3223 or support@hmco.com

Table of Atomic Masses

For your reference

Element	Symbol	Atomic Number	Atomic Mass	Element	Symbol	Atomic Number	Atomic Mass
Actinium	Ac	89	(227)†	Gold	Au	79	197.0
Aluminum	Al	13	26.98	Hafnium	Hf	72	178.5
Americium	Am	95	(243)	Hassium	Hs	108	(265)
Antimony	Sb	51	121.8	Helium	He	2	4.003
Argon	Ar	18	39.95	Holmium	Ho	67	164.9
Arsenic	As	33	74.92	Hydrogen	H	1	1.008
Astatine	At	85	(210)	Indium	In	49	114.8
Barium	Ba	56	137.3	Iodine	I	53	126.9
Berkelium	Bk	97	(247)	Iridium	Ir	77	192.2
Beryllium	Be	4	9.012	Iron	Fe	26	55.85
Bismuth	Bi	83	209.0	Krypton	Kr	36	83.80
Bohrium	Bh	107	(264)	Lanthanum	La	57	138.9
Boron	B	5	10.81	Lawrencium	Lr	103	(260)
Bromine	Br	35	79.90	Lead	Pb	82	207.2
Cadmium	Cd	48	112.4	Lithium	Li	3	6.941
Calcium	Ca	20	40.08	Lutetium	Lu	71	175.0
Californium	Cf	98	(251)	Magnesium	Mg	12	24.31
Carbon	C	6	12.01	Manganese	Mn	25	54.94
Cerium	Ce	58	140.1	Meitnerium	Mt	109	(268)
Cesium	Cs	55	132.9	Mendelevium	Md	101	(258)
Chlorine	Cl	17	35.45	Mercury	Hg	80	200.6
Chromium	Cr	24	52.00	Molybdenum	Mo	42	95.94
Cobalt	Co	27	58.93	Neodymium	Nd	60	144.2
Copper	Cu	29	63.55	Neon	Ne	10	20.18
Curium	Cm	96	(247)	Neptunium	Np	93	(237)
Darmstadtium	Ds	110	(281)	Nickel	Ni	28	58.69
Dubnium	Db	105	(262)	Niobium	Nb	41	92.91
Dysprosium	Dy	66	162.5	Nitrogen	N	7	14.01
Einsteinium	Es	99	(252)	Nobelium	No	102	(259)
Erbium	Er	68	167.3	Osmium	Os	76	190.2
Europium	Eu	63	152.0	Oxygen	O	8	16.00
Fermium	Fm	100	(257)	Palladium	Pd	46	106.4
Fluorine	F	9	19.00	Phosphorus	P	15	30.97
Francium	Fr	87	(223)	Platinum	Pt	78	195.1
Gadolinium	Gd	64	157.3	Plutonium	Pu	94	(244)
Gallium	Ga	31	69.72	Polonium	Po	84	(209)
Germanium	Ge	32	72.59	Potassium	K	19	(39.10)
				Praseodymium	Pr	59	140.9
				Promethium	Pm	61	(145)
				Protactinium	Pa	91	(231)
				Radium	Ra	88	226
				Radon	Rn	86	(222)
				Rhenium	Re	75	186.2
				Rhodium	Rh	45	102.9
				Rubidium	Rb	37	85.47
				Ruthenium	Ru	44	101.1
				Rutherfordium	Rf	104	(261)
				Samarium	Sm	62	150.4
				Scandium	Sc	21	44.96
				Seaborgium	Sg	106	(263)
				Selenium	Se	34	78.96
				Silicon	Si	14	28.09
				Silver	Ag	47	107.9
				Sodium	Na	11	22.99
				Strontium	Sr	38	87.62
				Sulfur	S	16	32.07
				Tantalum	Ta	73	180.9
				Technetium	Tc	43	(98)
				Tellurium	Te	52	127.6
				Terbium	Tb	65	158.9
				Thallium	Tl	81	204.4
				Thorium	Th	90	232.0
				Thulium	Tm	69	168.9
				Tin	Sn	50	118.7
				Titanium	Ti	22	47.88
				Tungsten	W	74	183.9
				Uranium	U	92	238.0
				Vanadium	V	23	50.94
				Xenon	Xe	54	131.3
				Ytterbium	Yb	70	173.0
				Yttrium	Y	39	88.91
				Zinc	Zn	30	65.38
				Zirconium	Zr	40	91.22

*The values given here are to four significant figures where possible. †A value given in parentheses denotes the mass of the longest-lived isotope.

Visit: college.hmco.com/PIC/ebbingessentials2e For Technical Support: (800) 732-3223 or support@hmco.com

Print Supplements Available Separately

Student Solutions Manual (ISBN 0-618-49178-3)

David Bookin, *Mount San Jacinto College,* Darrell D. Ebbing, *Wayne State University,* Steven D. Gammon, *Western Washington University,* and Ronald O. Ragsdale, *University of Utah*

This manual contains detailed solutions to the in-chapter exercises and odd-numbered practice problems, general problems, and cumulative-skills problems. It also contains answers to the Review Questions. An independent reviewer has checked all solutions to ensure accuracy.

Study Guide (ISBN 0-618-49177-5)

Larry K. Krannich, *University of Alabama at Birmingham*

Each chapter of the *Study Guide* reinforces students' understanding of concepts and operational skills presented in the text. It includes the following features for each chapter: a list of key terms and their definitions, a diagnostic test with answers, a summary of major concepts and operational skills, additional practice problems and their solutions, and a chapter post-test with answers.

To purchase these supplements, ask at your bookstore, visit our website at college.hmco.com, *or call Houghton Mifflin Customer Service at (800) 225-1464.*

Visit: college.hmco.com/PIC/ebbingessentials2e   For Technical Support: (800) 732-3223 or support@hmco.com

System Requirements

Eduspace® System Requirements

Minimum Requirements	Strongly Recommended Hardware/Software
• Microsoft Windows 2000 or Windows XP *or* Macintosh OS* 10.1, 10.2, or 10.3	• Microsoft Windows 2000 with Microsoft Internet Explorer 6.0 or Netscape Navigator 7.1** *or* Microsoft Windows XP with Internet Explorer 6.0 or Netscape Navigator 7.1** *or* Macintosh OS 10.3 with Safari 1.2 or Netscape Navigator 7.0 (or higher)**
• Microsoft Internet Explorer 5.5 (or above) *or* Netscape Navigator 6.2 (or higher, 7.1 or higher for Windows XP or Macintosh OS 10.2)	
• Sun Java Run-time Environment (JRE) 1.4.x in your Web browser (note use of Microsoft JVM is **not supported**)	• Adobe Acrobat 6.0
• Acrobat Reader 4.0	• Display with thousands of colors and 1024 x 768 resolution or higher
• Display with 256 colors and 800 x 600 resolution or higher	
• 56 K modem connection	• T1 line, cable modem, DSL, or company Internet LAN

*Use of Macintosh OS 10.1 or higher is required for users of Math and Chemistry courses. Users of other courses may be able to successfully use Macintosh OS 9.2 with their Eduspace courses.

**Users of Math and Chemistry courses are strongly advised to use Microsoft Internet Explorer or Safari to ensure that symbols in exercises and test questions appear properly.

Depending on your course content, you may need one or more of the following plug-ins: Flash Player 7 or higher, QuickTime 5.0 (6 recommended), RealPlayer 9.0 (10.0 recommended), Shockwave 8.5.1.

What if I have Windows 98, Windows NT or AOL?
Although Houghton Mifflin has not performed certification of Eduspace on Windows 98, Windows NT, or AOL and therefore cannot say that we officially support those platforms, Houghton Mifflin is aware of users who are successfully using Eduspace in those environments. If you encounter problems, once you are online with AOL, start an additional browser such as Microsoft Internet Explorer and access Eduspace using that browser. Note that when using any browser with Eduspace, cookies must be enabled.

Visit: **college.hmco.com/PIC/ebbingessentials2e**   For Technical Support: **(800) 732-3223** or **support@hmco.com**